UNIFORMS ILLUSTRATED NO. 6

1. Two US soldiers in DPM combat clothing testing the new 'Stinger' surface-to-air missile. No rank or regimental badges are visible. (US Army)

UNIFORMS ILLUSTRATED NO. 6

NATO Uniforms Today

**Belgium ● Canada ● Denmark ● France ● West Germany
Greece ● Italy ● Luxembourg ● Netherlands ● Norway ● Portugal
Spain ● Turkey ● United Kingdom ● United States**

DIGBY SMITH

a&ap

ARMS AND ARMOUR PRESS
London—Melbourne—Harrisburg, Pa.—Cape Town

Introduction

Uniforms Illustrated 6: NATO Uniforms Today
Published in 1984 by Arms and Armour Press, Lionel Leventhal Limited, 2–6 Hampstead High Street, London NW3 1QQ; 11 Munro Street, Port Melbourne 3207, Australia; Sanso Centre, 8 Adderley Street, P.O. Box 94, Cape Town 8000; Cameron and Kelker Streets, P.O. Box 1831, Harrisburg, Pennsylvania 17105, USA

British Library Cataloguing in Publication Data
Smith, Digby
NATO uniforms today. – (Uniforms illustrated; 6)
1. Uniforms, Military – Pictorial works
I. Title II. Series
355.1′4′091821 UC480
ISBN 0-85368-665-3

Editing and layout by Roger Chesneau.
Typeset by CCC, printed and bound in Great Britain by William Clowes Limited, Beccles and London.

In this brief volume we have tried to present as many different uniforms, together with details of badges of rank, regiment and formation, as possible. Inevitably, the result is a compromise between our objective and the material available.

Over the last thirty years there has been a gradual convergence in styles and colour of combat dress and equipment, with olive drab predominating when camouflage material (or Disruptive Patterned Material – DPM) is not worn. The many variants of camouflage patterns in use reflect the many different approaches to concealment in the field. The German Army, however, maintains plain olive drab for all forms of combat and fatigue clothing.

Armies deploying mountain troops or that have operational commitments in the Far North (Norway, Britain, Germany, Italy, France and Spain) clothe their troops in such environments with the universal, white, hooded smock, trousers and gloves, with weapons being masked by white tape.

Although badges on combat clothing are usually limited to the essential minimum and are of a black or other dull colour, the American Army still manages to sport a wide range of emblems – as will be seen.

Service Dress presents a slightly more colourful aspect and permits national traditions to be exhibited. Most armies prefer a shade of khaki for this uniform, but the German Army has worn light grey since the formation of the *Bundeswehr* in 1956 – a definite break with the *Feldgrau* of the *Wehrmacht*, but retaining a tenuous traditional link.

This book is intended to present the 'compleat' soldier, so while badges are to be seen on many of the plates, they are of secondary importance.

Recent research into ballistic protection has led to the introduction of reinforced plastic composition helmets in some armies (the British and American, for example), but others cling to the old, steel, US-style 'NATO–helmet'.

In almost all cases, combat web equipment is olive drab. The British Army uses butyl-nylon instead of canvas for this item because it is easier to decontaminate. Clothing to protect the soldier from nuclear, biological and chemical agents (NBC) is almost universally olive drab and is, incidentally, cordially hated for the acute discomfort it inflicts.

We are most grateful to the various Military Attachés who have been most generous in their provision of photographs, and special thanks are due, once again, to the staff of *Soldier* Magazine who leapt into the breach to help us out at the eleventh hour.

Digby Smith, January 1984

◀2

2. Rifleman of the Royal Irish Rangers. With the DPM combat smock and olive-green woollen jersey, this man wears the dark-green ('Ranger' green) bonnet and hackle and the silver crowned harp badge of his regiment. He is firing the Sterling submachine-carbine. (*Soldier* Magazine)

3. NATO gathering. Left to right: Major, VIII Canadian Hussars, in olive-green combat dress with black beret; German driver (in 'jeep'); American major of airborne troops (note airborne shoulder title and paratrooper's wings on chest) in olive-green combat kit; Major, Royal Anglians, in DPM temperate combat kit with regimental beret in khaki with black patch behind the badge. (Author's Collection)

4. NATO Artillerymen – a mixed bag at a Lance firing session in Germany, April 1983. Those in DPM camouflage jackets, with brass cannon cap badge, are from the British Royal Artillery; on the right is a captain of the *Bundeswehr* artillery, with red beret and bronze crossed cannons in a wreath as a badge; and on the left is a sergeant of the Royal Dutch Artillery. (*Soldier* Magazine)

▲3 ▼4

5▲ 6▼

5. A major of the Belgian Commandos instructing a class prior to an exercise. Note the black beret with brass regimental badge, gold bar and star rank badges under the regimental badge on the shoulder, black and white commando arm badges and paratrooper's wings. (Belgian Army)
6. Premier Sergeant-Major, military police of the 1st Belgian Corps, in black motorcycle leathers. The five silver chevrons of his rank are worn on black backing. (Belgian Army)

▲7 ▼8

7. Farewell parade of Colonel Demanet. At the extreme right of the picture can be seen an *adjudant* (master sergeant) with one, six-pointed silver star and next to him a *corporal* (lance-corporal) with two silver chevrons. (Belgian Army)

8. Two Belgian *adjudants* (warrant officers, 2nd Class) and a lieutenant in khaki Service Dress, viewing the British 'Wheelbarrow' remote-controlled bomb-disposal device. The rank stars are silver for warrant officers and gold for officers, and are worn on pentagonal collar patches in the regimental facing colour. Shoes are brown, badges and buttons gold. The lieutenant wears the traditional 'B' paratrooper's badge on his right arm. Above the regimental cap badge is the black, yellow and red national cockade. (Author's Collection)

9. Private, Royal Canadian Dragoons, wearing black beret and olive drab combat shirt. Note angled breast pockets and regimental cap badge and shoulder titles. (Canadian Army)

10. Military Policeman of the Canadian Army on point duty during exercises in southern Germany. The shoulder title on his olive drab combat shirt reads 1 MP PL (his unit designation) in light khaki on olive drab. (Canadian Army)

11. A unit of the Canadian army in Service Dress during a march past on Armistice Day. Uniforms and berets are dark green, buttons and rank badges gold; note the poppies worn next to the cap badges. Rank badges follow the British Army scheme for enlisted men, but officers have worn a tri-service system of gold bands on cuff (and on the shoulder when in shirt-sleeves or combat dress) for the last ten years. The lieutenant commanding the contingent wears specialist paratrooper's wings on his left breast and his formation crest on the right breast pocket. Gloves and shoes are black. The second corporal in the nearside rank wears US-pattern paratrooper's wings. (Canadian Army)

9▲

10▲ 11▼

▲12

▲13

◀14

12. Canadian infantryman in olive drab combat gear during an exercise in southern Germany. Note the training attachment in the muzzle of the 7.62mm SLR. (Canadian Army)

13. Sergeant McDougall, Royal Canadian Parachute Regiment, during an exercise in southern Germany. The rank stripes and shoulder title are in light khaki on olive drab; the national flag is in red and white. (Author's Collection)

14. A corporal of 4 Field Squadron, Royal Canadian Engineers, operating an anti-tank projectile launcher. (Canadian Army)

15. A Danish *Seniorsergent 2* (equatable with a British Army warrant officer) of the *Heeresmateriel-Kommander* (Ordnance Corps) as shown by the gold grenade and wreath badge on the dark-blue beret. The fatigue smock, jersey and shirt are olive green, the rank chevrons silver. (Danish Army)

16. A Danish sergeant in olive drab combat dress. The weapon is the Hovea 9mm M49 machine carbine, a derivative of the Swedish Carl Gustav weapon. Magazines for 30 or for 50 rounds are available. (Danish Army)

17. Danish Women's Corps members undergoing NBC training. Note the US-style helmets, the butyl-nylon ammunition pouches and the respirators ready for use (the two women nearest the camera). (Danish Army)

15▲

16▲ 17▼

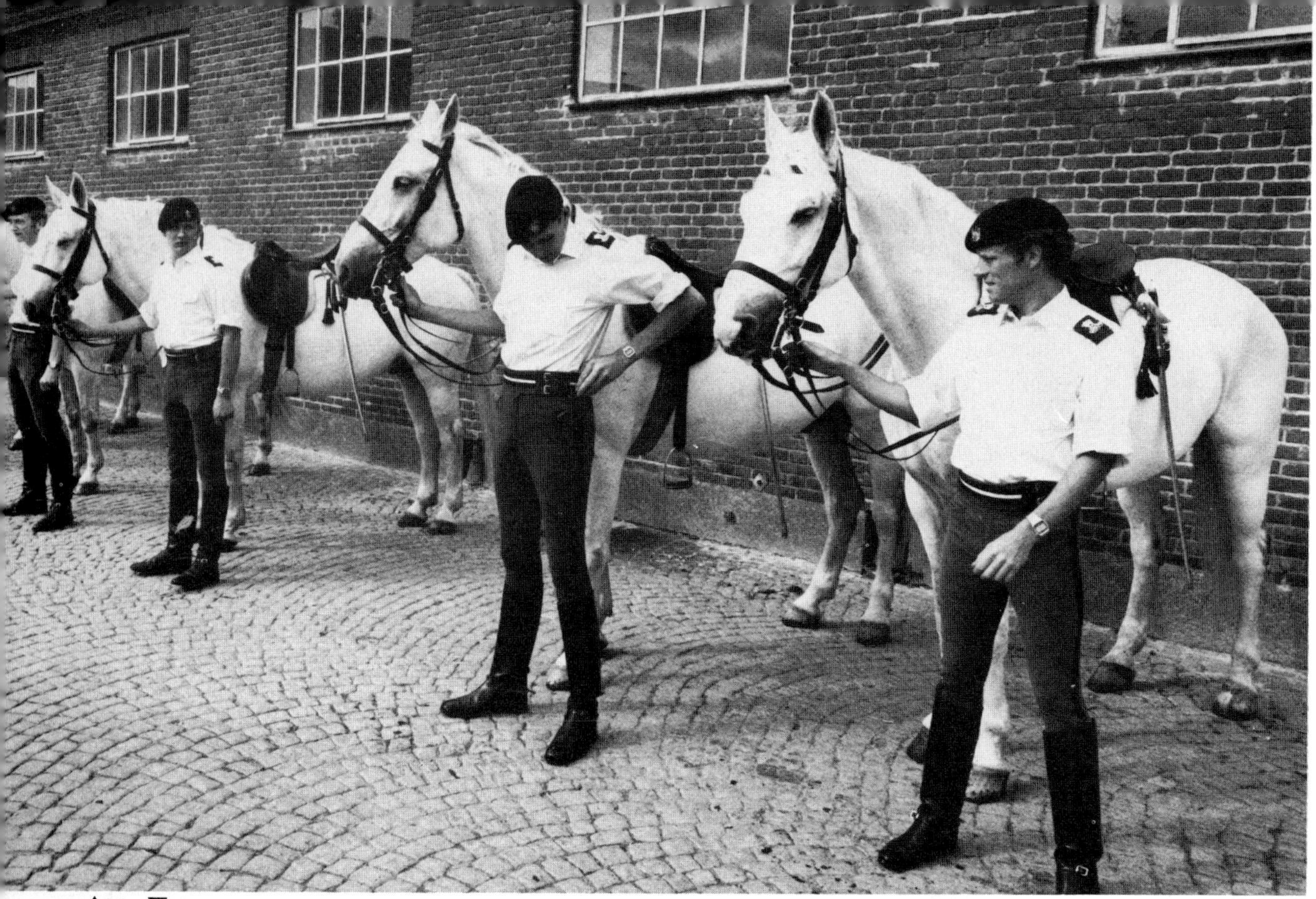

▲18 ▼19

18. Men of the mounted squadron, Royal Danish Hussars, in summer stable dress. The badge on the dark-blue beret is a diving eagle and scroll; on the shoulders is the Danish royal cypher; the breeches are khaki. (*Soldier* Magazine)
19. Lieutenant of the Danish *Leibgarde* Regiment in dark-green 'tent hat' and Service Dress jacket. The hat braid, piping and tassel are silver, the cockade silver and red. The white backing to the collar badges denotes the 2nd Battalion. On chest and upper arm are skill-at-arms badges. (*Soldier* Magazine)
20. Crewmen of a French Army Panhard 8 × 8 armoured car, in olive-brown coveralls with olive-brown armoured vehicle crewman's helmets with built-in intercom sets. The infantryman (in dark-blue beret) in the background is from an unidentified unit. (Author's Collection)
21. Troops of the II French Army Corps pass German soldiers in an end-of-exercise parade in southern Germany. The berets are dark-blue, the uniforms olive-brown and the corps badge red, white and blue. (Author's Collection)

20▲ 21▼

▲22

▲23

◀24

22. The one-armed Commanding General of the II French Corps is received in the headquarters of the German II Corps in southern Germany. The four silver stars of his rank can be seen on his dark-blue beret and on the dark-blue slip-ons on the shoulder-straps of his olive-brown combat dress. (Author's Collection)
23. A French tank crewman snatches a bite during an exercise. All clothing is olive drab, leather equipment brown. (II German Corps)
24. A class at the Federal German Command and Staff College, Hamburg. Front row (left to right): US Lieutenant-General Blanchard, in green Service Dress wearing the black and gold arm patch of the 1st Air Cavalry Division (his last combat posting); a German *Brigade-General*, with the gold buttons, star and wreath of his rank, and with a general's ornate gold and red collar patches; a colonel of the US Army; *Oberst* (Colonel) H. Hagenbrück of the Federal Army. This officer wears the distinctive crimson and silver collar patches of the German General Staff Corps, the three silver stars and wreath of his rank and parachutist's wings over his right pocket; a major of Britain's Royal Army Ordnance Corps, in khaki Service Dress with gold buttons, rank crowns and regimental collar badges; shirt and tie are light khaki, shoes brown. In the background are five officers of the German Army in light-blue shirts, black ties, grey epaulettes showing rank and regiment (by the coloured piping) and black trousers and shoes. (Author's Collection)
25. *Gebirgsjäger* (mountain troops) of the *Bundeswehr (Heer)* (Federal Army) in parade uniform. Cap and blouse are light grey, trousers and leatherwork black. Collar patches are silver on green, the divisional arm patch shows an edelweiss on a green field within a border whose colour indicates the wearer's division or brigade. On the left side of the cap is worn a bronze edelweiss badge of the same design as that used in the *Wehrmacht* during the Second World War. The cap badge features the black-red-yellow national cockade over crossed bronze swords. The officer is wearing the golden eagle in a wreath, on the left breast pocket, indicating top performance in the annual battle-fitness training. Two of the men are wearing silver stars on the left forearm; they are candidates for officer training. The drum hoops are black, red and yellow. (Author's Collection)
26. German officers with a Turkish Army colonel. The Germans (from the left three captains, a colonel and a major) wear Service Dress with the *Schiffchen* (forage-cap) bearing the national cockade, black leather waistbelt with silver buckle plate, bearing the eagle in a wreath, and black boots. The major wears grey suede gloves. The armpatch is that of the Command and Staff college – black over red over gold, in the centre the black eagle on a gold ground. The Turkish Army Colonel wears khaki working uniform with gold badges, chinstrap and peak embroidery (oak leaves), crimson collar patches with gold buttons (General Staff) and the gold wreath, crescent and stars of his rank on a crimson background. His shirt is light-tan, tie and shoes black. For formal occasions the battledress top would be replaced by a khaki Service Dress tunic. (Author's Collection)

25▲ 26▼

27. Dispatch Riders of the Federal German Army take a break while on manoeuvres near the Czechoslovak border. The combat uniform is olive drab, as is the webbing equipment. (Author's Collection)
28. Federal German engineers during an assault crossing of the Danube. They wear olive drab combat clothing and life-jackets.
29. An infantry major of the Federal German Army, in olive drab combat uniform, during an exercise on the Danube. Note the camouflaged cover to the US-style helmet. The rank badges are in white embroidery. (Author's Collection)
30. A *Gebirgsjäger* major, in olive drab combat shirt, breeches and stockings, with the special ski-cum-mountaineering boots used by these troops, greets a visiting Pakistani major-general. He is wearing the light-grey Service Dress cap. (Author's Collection)
31. *Gebirgsjäger* of the Federal German Army loading a 105mm light howitzer onto mules. Logistic resupply in the Alps is increasingly being taken over by helicopters. (Author's Collection)

▲27 ▼28

29▲

30▲ 31▼

▲32 ▼33

34▲ 35▼

32. Federal German Army infantrymen waiting to move forward for an assault crossing of the Danube. Their *Marder* armoured personnel carrier holds their assault boat on an improvised transport and launching-frame. The man at the left is wearing the olive drab, fleece-lined parka. (Author's Collection)
33. The crew of a German *Marder* armoured personnel carrier watch a US anti-tank section marching past. The Germans are *Jäger* (Rifles), with dark-green berets bearing the gold oak leaf sprig badge of that arm. (Author's Collection)
34. A German lieutenant and a French liaison officer (rank badges worn on the front of the tunic and obscured here) report on progress from a wood during exercises near Ingolstadt. (Author's Collection)
35. A *Kampfschwimmer* (combat frogman) of German engineers prepares to enter the Danube near Ingolstadt to assist in recovering a sunken army ferry pontoon. Note his weighted boots. (Author's Collection)

▲36 ▼37

36. A 4-star German General (centre), with pale-grey *Schiffchen* (forage-cap) and olive drab parka bearing yellow rank badges, conferring with a Swiss *Oberstkorpskommandant* (Lieutenant-General) and an *Oberstdivisionär* (Major-General). Behind the General is a *Bundeswehr* colonel wearing the AFCENT pocket badge. (German Army)

37. *Fallschirmjäger* loading equipment into a *Luftwaffe* transport helicopter prior to embarking on a training mission. Note the 3-point helmet suspension. All uniform and equipment is olive drab except the sleeping-bag and the respirator case which are dark olive green. The NCO at the extreme right is a corporal (*Unteroffizier*), as shown by the white edging to his shoulder-straps. (Author's Collection)

38. A *Feldwebel* (Sergeant) of the Federal German *Feldjäger* (Military Police) in walking-out dress. The beret is red with its silver badge showing an eagle within an eight-pointed rayed star, all within an oak leaf wreath having the national flag at its base point. The coat is light grey and the collar patches silver on orange; the shoulder straps are edged orange and have a gold inner braid containing a silver chevron, point uppermost. On the right pocket is the eight-rayed silver star of the Military Police corps. (German Army)

39. A group of Federal German Army *Jäger* (Infantry) undergoing instruction on the MG2 light machine-gun. The berets are dark green with a gold badge in the form of three oak leaves within a corded ring, the national flag at the base. At left the instructor, an *Oberfeldwebel* (Staff Sergeant with two white chevrons within a white outline on his shoulders), watches a trainee clearing a blockage. (German Army)

38▶

39▼

40. Two *Gefreiter* (lance corporals) of the Federal German Army in olive drab fatigue dress. The man on the left wears the *Schiffchen* side hat in olive drab with a black, red and gold cockade; his *Jäger* (Infantry) companion with dark green beret and gold badge sports the bronze eagle pocket badge for combat fitness and two proficiency badges over his right pocket. (German Army)
41. A Greek paratrooper in camouflaged combat dress. The beret has the blue and white cockade which replaced the traditional, circular, crowned pattern worn until the monarchy was overthrown in 1974. On the left upper arm is the winged vertical sword badge of the parachute regiment bearing the motto OTOλ MONNIKA. Paratrooper's wings are worn on the left breast. Note how the lower sets of ammunition pouches are tied to the thighs. (Greek Army)
42. Corporal, Greek Paratroopers, in combat dress. This man wears the olive drab fatigues with NATO helmet. His badges of rank (two white chevrons) are shown on the collar, just above his paratrooper's wings. (Greek Army)
43. Greek Sergeant 1st Class, summer Service Dress. The olive-brown cap and uniform replaced the green clothing whose colour was almost identical with that used by the American Army up until 1974. The yellow on red rank badges consist of four chevrons, a flaming grenade and an overall arc. (Greek Army)
44. Greek lieutenant-colonel, winter Field Uniform. Shoulder rank badges consist of two, six-pointed gold stars under a golden flaming grenade. The regiment is indicated by the badges on the coloured collar patch, and on his left pocket this officer wears the badge of a staff college graduate. (Greek Army)

◀40

41▲ 42▼

43▲ 44▼

▲45

45. Greek infantry in summer fatigue/training uniform. This khaki battledress is modelled very closely on that worn by British troops during the Second World War. (*Soldier* Magazine)

46. An Italian *Alpini* (mountain troops) lieutenant-general in light-khaki summer Service Dress (shirt-sleeve order). The light-grey *Alpini* hat bears the gold on red eagle and wreath badge of a general officer, with his three gold stars on a silver label on the left side just above the gold boss into which the eagle feather is set. On his shoulder-straps are his rank badges – three golden stars above a band of golden embroidery. The silver star of Savoy can be seen on his collar. Behind him is his aide-de-camp, a lieutenant-colonel of *Alpini*. This officer wears the *Alpini* cap badge of an eagle over a hunting horn and crossed rifles, regimental collar patches with the silver star of Savoy, and rank badges of a mural crown under two gold stars. To the lieutenant-general's left is a lieutenant-colonel of the Federal German Army. (Author's Collection)

47. NATO officers line up. From left to right: Major, Dutch artillery; Major, British light infantry; Major, Greek cavalry; Major, Norwegian infantry; Major, Royal Army Ordnance Corps; Lieutenant-Colonel, Italian *Bersaglieri*; Major, German tank troops; Major, French marine infantry; Major, Belgian infantry; Colonel, Turkish General Staff. (Author's Collection)

48. Belgian Military Police: a woman corporal in olive green fatigues, with a private on train duty in Germany. Note flaming grenade cap badges. (Belgian Army)

46▶

47▲ 48▼

▲49

▲50 ▼51

52▲

53▲

49. A captain of the Danish *Norrejyske* Artillery Regiment in walking-out dress. Beret and collar badges are identical (in gold, a flaming grenade over crossed cannon barrels, on the body of the grenade a rampant lion over nine hearts, four over three over two). Over the right breast are skill-at-arms and qualification badges; on the shoulder are the three small, gold stars of a captain. The dark-green beret and tunic are worn with medium-grey trousers having a black side piping. (Danish Army)
50. Danish infantryman in NBC clothing. Note the side-mounted filter on the respirator which enables the field telephone to be used while wearing NBC protective clothing. Apparently, there are no overboots issued with this suit. (Danish Army)
51. French infantry debussing from the VAB 6-wheeled APC; they are carrying respirator pouches on their brown leather belts. (Author's Collection)
52. Federal German Army infantry in olive drab combat clothing and NATO helmets, and carrying the 7.62mm G3 SLR and the MG2 light machine-gun. (German Army)
53. *Kommandos* of the Royal Netherlands Army. These men are abseiling and carry UZ1 submachine-guns; on the right is a *Korporal* (lance-corporal in the British Army), recognizable by his single yellow rank chevron. Their headgear is the 'cap comforter', a sealed tube of khaki wool which doubles as a scarf. (Royal Netherlands Army)
54. Two privates, Royal Netherlands Army, in winter Combat Dress. The man nearest the camera is wearing a parka in addition to the usual olive drab tunic; he also has khaki woollen gloves. Both men are carrying the NATO SLR as used in the British Army. (Royal Netherlands Army)

◀54

55. Private, Royal Netherlands Army, in winter Combat Dress. Beneath the grass and netting can be seen the multi-coloured cloth cover of the helmet (which does not cover the shining helmet rim!). Details of the olive drab parka and webbing – with blackened metal fittings – can be seen clearly. (Royal Netherlands Army)
56. Two students at the Italian School of Artillery in summer Combat Dress. Note school shoulder-titles. The white, 5-pointed star of Savoy decorates both collar tips. (Author's Collection)
57. *Alpini.* These mountain troops are well known for their grey soft felt hats with eagle's feather trim. (Author's Collection)
58. Italian paratroops in training equipment. These men wear the brown and green camouflaged combat clothing with olive drab web equipment. Note fighting knife on left calf. (Author's Collection)

◀**55**

56▶

57▼

58▼

▲59

60▶

59. A lieutenant-colonel of the *Lancieri di Montebello* in shirt-sleeve order. Note the gold rank stars and mural crown and the enamel collar badges (green, with three points) with the silver star of Savoy. Over the left breast pocket is the golden eagle and wreath badge of the Italian Army Staff College. (*Soldier* Magazine)

60. Ceremonial parade of the *Lancieri di Montebello*. The cap badge is a flaming grenade on crossed lances. The officer's sash is light blue. Note that NCO's rank stripes are worn on the epaulettes. (*Soldier* Magazine)

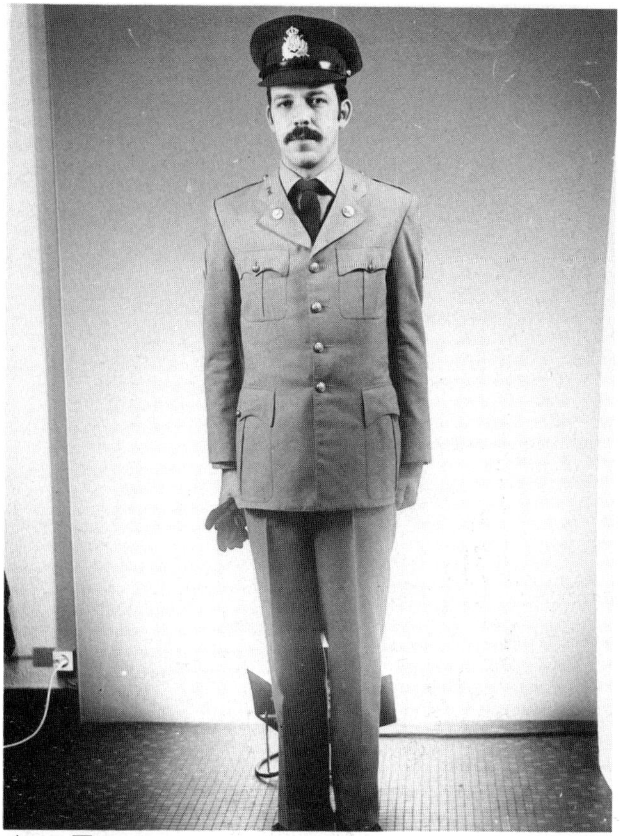

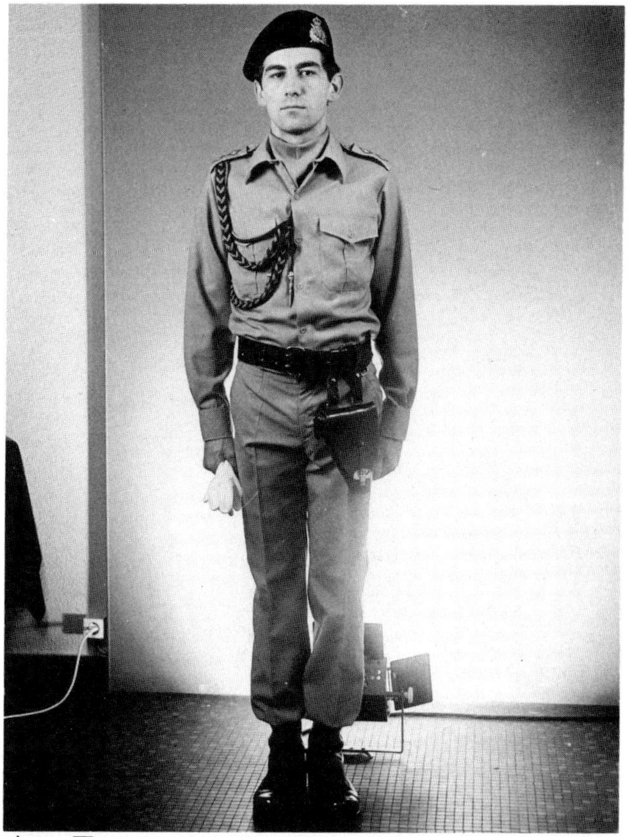

▲61 ▼62

▲63 ▼64

61. The uniform of Luxembourg's *Force Publique* reflects a mixture of British, American and Dutch influences. The gold cap badge worn by this *Adjudant* (Company Sergeant-Major) of Infantry, in summer Service Dress, is remarkably similar to that worn by the Royal Netherlands armies (both ruling houses originating from the lines of the tiny German mini-state of Nassau-Oranien), the general cut of the light-khaki uniform is British and the arrangement of the collar badges (the corps badge on a disc for other ranks, but worn 'clear' by officers) is American. Above the corps badge is the reversed grand ducal cypher (J) under a grand ducal cap. All badges and buttons are gold as are the three chevrons (point up) on both upper arms. (Ministère de la Force Publique)

62. Luxembourg *Volontair* (Private), summer Combat Dress. The American influence can be seen very clearly here from the camouflaged helmet cover to the LUXEMBOURG patch on the right chest, the name patch on the left and the red, white and blue national flag worn on the upper left arm. Clothing and webbing are olive green, the high boots are black. The standard 7.62mm NATO SLR is carried. (Ministère de la Force Publique)

63. *Lieutenant in Premier*, summer Parade Dress. All ranks of the *Force Publique* wear their ranks (in cloth) on the shoulder-straps in shirt-sleeve order; here we see the two, British-style 'pips' of a first lieutenant. The aiguilette, brown leather Sam Browne belt with pistol holster, light-tan cravat and white gloves are parade additions to the usual summer uniform. (Ministère de la Force Publique)

64. *Soldaat 1e. Klasse* (Private 1st Class), 2nd Battalion, Infantry Regiment 'Van Heutz', Royal Netherlands Army. Rank is shown by the single red chevron, point up, on the olive-khaki shoulder-strap slip-on. The beret is khaki and bears the regimental cap badge – an eight-pointed, rayed star on the usual lateral figure of eight brass surround – on a patch of cloth in the regimental colours. The light-tan shirt is worn with olive-khaki trousers, brown high boots and the increasingly ubiquitous khaki jersey originally introduced by the British Army and now adopted by many others. (Royal Netherlands Army)

65. *Soldaat 1e. Klasse* and Private, summer Fatigue/Combat Dress. For sentry duty the beret is replaced on this exercise by the US-pattern helmet. The junior NCO (right) wears a single red chevron on his shoulder-straps; on his right shoulder are his ear-defenders. All clothing and equipment is in dark olive drab. On his left breast pocket the private wears the badge of the formation in which he is currently serving. (Royal Netherlands Army)

65▼

▲66

▼67

66. Junior Ranks in Service Dress. The dark-brown jacket bears the following distinguishing badges: on the collar, the regimental badges, varying in design and colour for each different regiment or corps (here unidentified); on the shoulder-straps, the badges of rank (to the right is a *Korporal 1ᵉ Klasse* with two yellow chevrons); on the upper left arm, the Royal Netherlands' lion with motto: JE MAITENDRAI in orange thread; on the left pocket, the badge of the formation to which the man is currently assigned (it must be assumed that all but the *Korporal 1ᵉ Klasse* are unassigned and are probably recruits in training); above the left pocket, medals as appropriate. Buttons are gold. The trousers are light khaki, the boots brown. (Royal Netherlands Army)
67. HRH Prince Bernhard of the Netherlands, a Lieutenant-Colonel, RNA and Staff Sergeant R. Marritt, Royal Army Ordnance Corps. Although taken some time ago this photograph shows the current Netherlands and British Army Service Dress and the rank badges of a general and lieutenant-colonel in the former. HRH wears the then royal cypher above his golden rank stars. His collar badges are of the General Staff and his hat band is red. The large, three-button cuff flap is peculiar to Generals. Staff Sergeant Marritt wears the cap and collar badges, buttons and red and blue lanyard of his corps. (Sergeant-Majoor F. J. Bertens)

68▲

69▲

68. Private of Engineers, Barrack Dress. The khaki beret bears the silver corps badge on a blue, red-edged background; the jersey is khaki the shirt light-tan. (Royal Netherlands Army)

69. A lieutenant of the Dutch *Stoottroepen* Regiment in Service Dress. The gold beret badge (a sword within antlers and the lateral scroll bearing STOOTTROEPEN) is on a red patch on the light-khaki beret. All badges and buttons on the dark-brown jacket are gold, the collar patches red. The paratrooper's wings are white and light blue, the Netherlands lion arm badges orange. Shirt and tie are light tan, trousers medium brown, shoes brown. (*Soldier* Magazine)

70. A Dutch sergeant (one gold chevron on dark-brown shoulder-strap) of the Infantry Regiment watches a mechanic servicing an armoured personnel carrier. The beret is khaki, all other clothing olive drab. The gold cap badge is on a red backing. (*Soldier* Magazine)

70▼

71. Norwegian Private, School of Infantry, summer Combat Uniform. This man wears the standard NATO helmet (in metal), a brown, tan and green camouflage smock, olive-green webbing and olive-green trousers. The weapon is the West German MG2, an adaptation into 7.62mm calibre of the MG41 of Second World War fame. The S on the shoulder denotes 'School'.
72. Norwegian Army uniforms. From left to right: (i) Sergeant, Women's Services. The cap badge is the royal cypher OV (Olaf V); rank badges are worn on the collar; buttons and badges are silver, the clothing green/grey. (ii) Winter Combat Uniform. The olive drab combat smock and trousers are covered with a white, hooded snow jacket and trousers. Note that the black leather boots are designed to be worn as ski boots. (iii) Summer Combat Uniform. The same camouflaged top and olive-green trousers, with puttees, as shown before. (iv) Summer Working Dress. All items of clothing are in olive green; the cap badge is the silver, blue and red crowned cockade; the ski boots are worn throughout the year. (Norwegian Army)
73. A Portuguese staff sergeant and a Canadian private pictured during a NATO exercise. The staff sergeant is wearing olive drab fatigues and cap, with his rank (four golden chevrons) shown on the shoulder-straps. The Canadian has taped his field dressing to his webbing strap. (Portuguese Army)

◀**71** **72**▲ **73**▼

▲74 ▼75

74. The Portuguese Army's NBC clothing is olive drab in colour and consists of the long over-smock (with detachable hood), gloves and leggings worn over the usual combat gear. The respirator is of a type which makes use of the field telephone rather difficult. (Portuguese Army)
75. Two Portuguese engineers fix a detonating set to demolition charges on a viaduct. They wear the new US-type helmet with level bottom rim under the camouflage cover. The colouring of the combat suit is brown on light green and stone. (Portuguese Army)
76. Portuguese infantrymen in winter setting. Note the black, roll-neck pullover worn under the camouflage tunic. They carry the Spanish version of the NATO 7.62mm SLR also used by West Germany. The web equipment is olive drab. (Portuguese Army)
77. Official photograph of the current Portuguese Army combat suit. Note camouflaged shirt and bellows pockets on chest and thighs.
78. *Cabo* of the Portuguese Parachute Regiment in summer camouflage combat dress. The beret is dark green, the badge a silver eagle within a gold wreath. The camouflage pattern is light and dark green with brown patches, the webbing olive drab. On his shoulders are black epaulette loops with two light-blue chevrons. He carries the Spanish version of the 7.62mm SLR. (Portuguese Army)

▲76

77▼

78▼

▲79 ▼80

▲81 ▼82

79. A Spanish paratrooper in training equipment. All clothing is olive green, boots black, gloves brown. The gold and black army eagle badge is worn on the front of the parachute pack. (Author's Collection)
80. Private, Spanish Foreign Legion, in summer Combat Dress. The uniform is light olive green, hat tassel red, badge gold; equipment black leather with brass fittings. The brass belt buckle bears the army eagle badge. (Author's Collection)
81. *Cabo Primero* (senior corporal) of Spanish infantry, summer parade dress. The army eagle is worn on the brass cap and belt plate; buttons are brass, leatherwork black. The infantry badge (crossed sword and arquebus behind a hunting horn) is worn in gold on the red rhombic collar badges. The cap and uniform are khaki, the three rank stripes are red. (Author's Collection)
82. A major of the Turkish General Staff in battledress with field cap. The khaki uniform and cap are embellished with gold and crimson badges as follows: cap – oak leaves and crescent and star; shoulders – oak leaves, crescent and star plus two large five-pointed stars; collar – crimson patch (General Staff) with gold button.
83. A Turkish infantry company on parade. The uniforms and webbing are khaki, the national shoulder flash red and white as are the helmet decals. The pennant (at rear) is that of an infantry company.
84. A Turkish military band in summer parade uniform. The collar patches are light crimson with a golden lyre, the rank badges and national patch crimson and white as are the helmet decals. (*Soldier* Magazine)

83▲ 84▼

◀85

86▲

85. Corporal of Turkish Military Police, summer Combat Dress. The olive drab uniform is set off by the red and white armband and brown leather pouches. (*Soldier* Magazine)

86. Turkish Colonel (*Albay*) in light-tan summer shirt-sleeve order. The cap badge and shoulder badges are gold and crimson; note the gold oak leaf edging to cap peak. (*Soldier* Magazine)

▲87 ▼88

87. A Turkish Major-General (*Tümgeneral*) in summer Service Dress. The crimson collar tabs have golden oak leaf embroidery and button, and on the shoulders are two gold stars over a crimson and gold oak leaf wreath on crossed sabres. The shirt, jacket and trousers are light tan, the tie is black and the Sam Browne belt is of brown leather. (*Soldier* Magazine)

88. Two Turkish second lieutenants (*Teğmen*) of the Signals under training. The man in the background wears dark-khaki Service Dress with tan shirt and black tie, his companion wears khaki shirt-sleeve order. Both have the light-blue Signals' collar patches with golden lightning flashes; rank stars and buttons are gold. (NATO-South)

89. Norwegian Military Police. The British influence is marked, shown in the red cap tops and lanyards and the white 'webbing' equipment with black armband and white MP sign. On the left arm is worn the rhombic sign of the North Norway Brigade; a white reindeer on black and below it the narrow strip of red and blue ribbon denoting an artillery unit. The cap badge is the silver, red and blue cockade. (Norwegian Army)

90. Members of the Welsh Guards with the new 4.85mm SLR and Light Support Weapon. They wear the Guards' khaki beret with standard DPM camouflaged combat clothing. (Ministry of Defence)

89▲ 90▼

91. Norwegian infantry in winter Combat Dress. Only the khaki webbing and black boots are left exposed; everything else is covered in white. Note the method of holding both ski sticks in the left hand while using personal weapons. (Norwegian Army)

▲92 ▼93

92. Norwegian infantry debussing from an M113 tracked armoured personnel carrier (APC), this version being fitted with a turret. Note the Norwegian flag on the upper left arm. The weapons are the Spanish/German version of the 7.62mm SLR and the West German MG2 light machine-gun. (Norwegian Army)
93. Norwegian infantry in winter combat clothing with the American-produced TOW (tube launched, optically tracked, wire guided) anti-tank missile system. (Norwegian Army)
94. British Royal Artillerymen with the 105mm Light Gun. These Gunners wear DPM Temperate combat kit with regimental (dark-blue) berets. The man in the foreground wears the white cap cover and lanyard of a gunnery instructor at the School of Artillery. (*Soldier* Magazine)

94▼

▲95

◀96 97▶

95. Left to right: lance-corporal REME in coveralls; warrant officer 2nd Class (Artificer) REME, also in olive-green coveralls with dark-blue beret and brass crown-and-wreath wrist badge; lieutenant-colonel, Royal Tank Regiment, in No. 4 Dress (Warm Weather Service Dress). This officer wears the golden aiguilette of a military attaché and the black astrakhan, ceremonial beret with twin gold stripes, silver regimental badge and brown-red and green feather hackle.

96. A corporal, Royal Military Police, and a bearded *Gefreiter* (lance-corporal) of the *Gebirgsjäger*. The RMP corporal is wearing the red and blue No. 1 Dress cap with black peak and chinstrap, gold badge and buttons; khaki combat shirt, blue on red MP brassard, khaki Service Dress trousers and black boots. The *Gefreiter* is wearing olive drab combat clothing with the single, diagonal white bar of his rank on the shoulder straps. (Author's Collection)

97. Rear view of the British dark olive green 1958 Mark II Pattern Web Equipment. This equipment was developed in a nylon-butyl material designed to be robust, water-resistant (non-water absorbing to obviate weight increase in wet weather) and shrink-resistant. It is also more resistant to chemical warfare agents than its predecessor and can be more easily decontaminated. (Directorate of Clothing & Textiles)

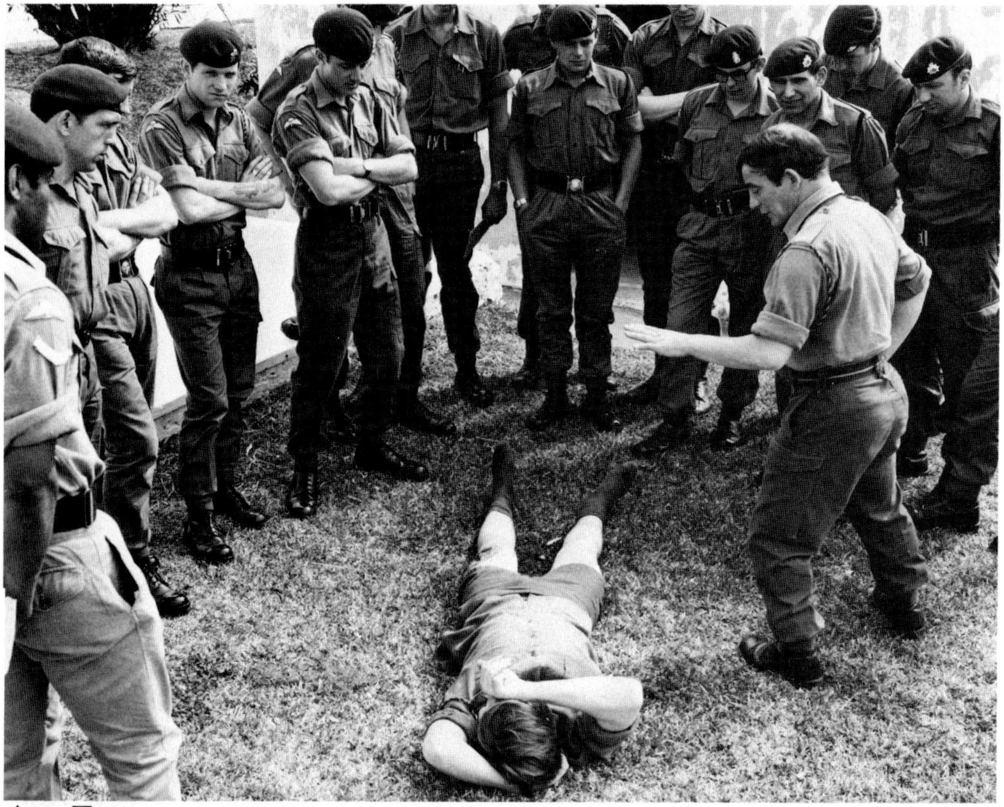

98. Members of 1 Parachute Logistics Regiment. These men wear the red beret of the Parachute Regiment to which they are attached, together with their own regimental badge, lanyard and stable belt; combat shirt and light-weight, olive-green trousers, puttees and combat boots. All wear paratrooper's wings. (*Soldier* Magazine)
99. Members of the Royal Army Ordnance Corps in Cold Weather Clothing. This picture shows the winter DPM combat clothing complete with winter warfare cap, gloves and boots. The driver of the Eager Beaver rough terrain fork-lift truck wears an olive-green crash-helmet. (Army Public Relations)

▲98 ▼99

100. An Ammunition Technician of the Royal Army Ordnance Corps in the Explosive Ordnance Disposal (EOD) protective suit. This cumbersome clothing contains glass reinforced plastic chest and pelvic armoured plates within the olive-drab covered Kevlar flexible armour. Note the two-way radio on his back. This equipment is used when tackling suspect explosive devices. (Author's Collection)

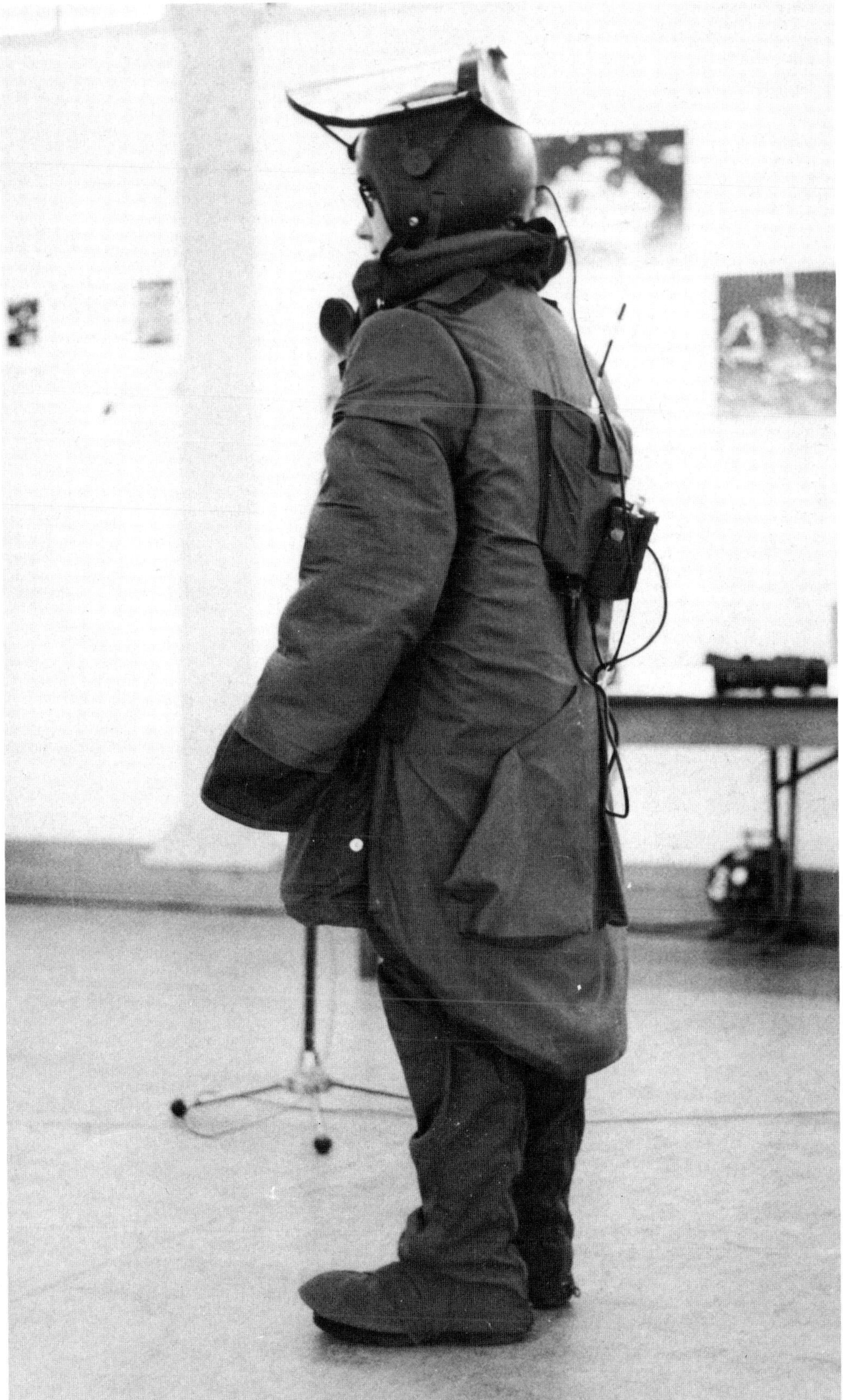

100▶

101. A British soldier tries the G3 version of the NATO SLR watched by two Portuguese instructors. (*Soldier* Magazine)
102. Major-General Barry Lane (VQMG and Colonel of the Light Infantry) inspects a guard of honour of his regiment. Note the dark-green regimental chevrons on buff backing and the black belts – a light infantry characteristic. The silver cap badge on the rifle-green beret is repeated (with a scroll) as a shoulder badge. (*Soldier* Magazine)
103. Men of the 10th Princess Mary's Own Gurkha Rifles marching past in Princes Street, Edinburgh in Service Dress. As riflemen, they wear black badges, buttons, gloves, shoes and belts. (*Soldier* Magazine)

▼101

102▲ 103▼

▲104

105▶

104. Warrant Officer, 2nd Class David Burrows, Army Air Corps, in Service Dress. Note the silver eagle collar badge, army pilot's wings and rank badge on cuff. (*Soldier* Magazine)

105. Three drill instructors of the Grenadier Guards (two sergeants and a staff sergeant) present their pace sticks for inspection by a regimental sergeant-major (RSM) of the Welsh Guards. Note regimental cyphers on the shoulder-straps, red sashes and brass peak edging, which varies according to rank, and the RSM's Sam Browne belt. The Grenadier Guards' cap headband is red. (*Soldier* Magazine)

▲107 ▼108

106. (Previous spread) Her Majesty the Queen inspecting a guard of honour of the Royal Regiment of Fusiliers. This regiment wears the traditional fusiliers' red-over-white, cut-feather hackle. The cap and collar badges are flaming grenades. Note the red sash worn by the Company Sergeant-Major on the right of the front rank. (*Soldier* Magazine)
107. Soldiers of the 1st Battalion, the Royal Scots, wearing the DPM combat smock, olive green light-weight trousers and the khaki regimental bonnet. (*Soldier* Magazine)
108. General Alexander Haig (later Secretary of State in President Reagan's administration until his resignation in mid 1982) tries on a *Bundeswehr Fallschirmjäger's* maroon beret, with its diving eagle badge, during a joint NATO exercise in Germany. On his left breast he wears the infantry combat badge in black on olive drab; on his collar are four black stars. The donor of the beret wears both *Bundeswehr* and French paratrooper's wings on his olive drab, summer-weight combat tunic. (Author's Collection)
109. A parade at Heidelberg in 1980 shows both the green US Service Dress (here worn by no fewer than six American Army Generals – their status indicated by cap peak embroidery, rank stars on the shoulders, extra-wide black cuff lace and double black stripes on their trousers and by the absence of arm-of-service collar badges) and the ceremonial uniform of dark-blue jacket and light-blue trousers with gold embellishments. (US Army)

109▼

110. A member of the crew of an American M60 A3 tank takes a break during Exercise 'Reforger 82' in Western Germany. Note the black on olive divisional patch on the left arm, the absence of a combat zone divisional patch on the right arm and the barely-visible, brown leather harness of the shoulder holster. (US Army)
111. The clothing worn by a soldier to protect him from the effects of nuclear, biological and chemical (NBC) agents imposes great strain and many limitations on the wearer. Combat efficiency is inevitably much reduced when wearing the NBC suit, particularly if ambient temperatures are high. NATO forces do not place as much emphasis on NBC training as do the armies of the Warsaw Pact states. This unidentified soldier of the 8th US Infantry Division is shown attempting to use a field telephone during Exercise 'Yellow Cloud' (a NATO NBC exercise) at Bruchsal in West Germany. His respirator is covered by an NBC-proof hood. (US Army)

▼110

111▶

STINER
RANGER
AIRBORNE

◀112 113▲

◀114

112. US Brigadier-General Carl W. Stiner, Assistant Division Commander of Operations, 82nd Airborne Division, pictured during Exercise 'Reforger 82' in West Germany. The helmet strap with name is a relatively new idea; the black on olive rank stars show up clearly on the tan, green, brown and black of the camouflage jacket. Above the black and olive 82nd Division patch (normal colours: white AA, blue circle, red background), Stiner wears the 'Ranger' and 'Airborne' titles which on Service Dress are in gold on green. Note that the helmet is held to the head by only two anchor points; this has been superseded by a three or four point system in many other armies. (US Army)
113. Three of these GIs are wearing olive drab, zip-fronted waterproof smocks while their black companion is wearing a buttoned olive parka with a balaclava-type hood under his helmet. Two of the men are carrying their first aid dressings in canvas pouches on the left-hand shoulder-brace of their webbing harness. No identifying badges are visible. (US Army)
114. Specialist Class 4 (Sp/4, a grade of private soldier) Rummel and a comrade of Company C, 1st Battalion, 48th Infantry, 3rd US Armored Division, pictured during training. Note the attachments in the muzzles of the M16 carbines which permit the weapons to operate in the automatic mode even though blank rounds are being fired. The Sp/4 badge is worn on the helmet front; waterproof, an olive drab smock covers the camouflage jacket. (US Army)

▲115 ▼116

115. An American PFC (Private First Class) during winter combat training in West Germany. The webbing pouches are for the M16 carbine magazines; on his right shoulder-strap can be seen the container for his ear defenders. (US Army)
116. US Sergeant Lee Barela of 313th Intelligence Company, 82nd Airborne Division, in summer camouflaged combat uniform. The rank chevrons are worn on the collar, but seem to have fallen off the helmet. Note the blackened metal clips on the ammunition pouches. His companion shows the Airborne title above the divisional patch, both in black on olive. (US Army)
117. US Captain J. Jackson, Commander of Company C, 1st Battalion, 3rd Armored Division, together with his radio operator (PFC Edwards) and runner during winter training in West Germany, April 1983. Olive drab anoraks cover the olive drab woollen jersey and camouflage tunic. The black leather combat gloves have reinforcing across the knuckles. Olive drab and black rank badges, Ranger title and divisional patch are worn; the web equipment obscures other details. (US Army)

117▶

▲118

118. Members of the 1st US Infantry Division at Ramstein air base (West Germany) after having been air-lifted in from Fort Riley, Kansas during Exercise 'Reforger 82'. They wear the summer-weight camouflaged combat uniform with olive drab webbing equipment, including a Rucksack-like backpack. Note the L-shaped small pack on the left hip, with a cut-out to accommodate the water-bottle. The black-on-olive divisional patch is the famous 'Big Red One'. Note the M16 with grenade launcher. (US Army)

119. Lunch break for members of a US patrol along the German border with Czechoslovakia. Note the rank badges worn on the black beret next to the regimental badge and the divisional patch. The Corporal (centre) wears the infantry combat badges on the left breast and the patch of his last combat formation on his left arm.

120. American soldiers of the 48th Infantry, 3rd Armored Division training for urban combat. Note captain's rank badges (fourth from left) on helmet and shoulder-straps, and the AIRBORNE legend over the divisional triangular badge. (US Army)

121. A US soldier of the 8th Infantry Division decontaminating his M113 APC. (US Army)

▼119

120▲ 121▼

122. This US soldier, wearing olive drab fatigues under his NBC clothing, is shown being stripped of his 'contaminated' NBC suit during a simulated decontamination exercise ('Yellow Cloud') in West Germany. Under the NBC hood, which is fitted with devices to permit use of a field telephone, he wears a black respirator. The black rubber gloves have cotton inner gloves and the NBC suit has a charcoal lining. (US Army)
123. US Army troops from I-Troop, 3rd Squadron, 2nd Armored Cavalry Regiment, liaise with West German Border Police near Rotz on the East-West border in southern Germany. All the Americans are wearing olive drab waterproof jackets over their camouflaged combat suits, black leather combat gloves and boots, camouflaged caps with black rank badges on the front, and 'skeleton order' web equipment with hip-pack. (US Army)
124. Members of the 3rd/11th US Armored Cavalry Regiment at Augustdorf, West Germany in May 1983, during the annual Böslager Cavalry Competition. Camouflage combat suits and caps are worn over olive drab T-shirts. On the left arm is the black and olive divisional patch; rank badges are worn on collar and cap fronts. The officer (light-coloured cap bar) is either a 1st or 2nd Lieutenant. On the extreme left can be seen a soldier of the Royal Canadian Dragoons, wearing black beret and leaping stag badge. (US Army)

◀**122**

123▲ 124▼

▲125

125. Men of the US 7th Infantry Divisional Artillery, with an immobilised self-propelled howitzer in Germany. The man in the foreground wears the patch of his last combat formation (four stars on a shield) on the right arm. (Author's Collection)

126. Signals specialist of the US 1st Armored Division, in olive green fatigues, setting up a mobile radio station. Note triangular divisional patch on left arm and rank badge on cap. (II German Corps)

▼126